U.S. Department of Transportation

Federal Railroad Administration

Low-Cost Warning Device Industry Assessment

Office of Research
and Development
Washington, DC 20590

Safety of Highway Railroad Grade Crossings

DOT/FRA/ORD-10/06

Final Report
July 2010

NOTICE

This document is disseminated under the sponsorship of the Department of Transportation in the interest of information exchange. The United States Government assumes no liability for its contents or use thereof.

NOTICE

The United States Government does not endorse products or manufacturers. Trade or manufacturers' names appear herein solely because they are considered essential to the objective of this report.

REPORT DOCUMENTATION PAGE			Form Approved OMB No. 0704-0188
Public reporting burden for this collection of information is estimated to average 1 hour per response, including the time for reviewing instructions, searching existing data sources, gathering and maintaining the data needed, and completing and reviewing the collection of information. Send comments regarding this burden estimate or any other aspect of this collection of information, including suggestions for reducing this burden, to Washington Headquarters Services, Directorate for Information Operations and Reports, 1215 Jefferson Davis Highway, Suite 1204, Arlington, VA 22202-4302, and to the Office of Management and Budget, Paperwork Reduction Project (0704-0188), Washington, DC 20503			
1. AGENCY USE ONLY (Leave blank)	2. REPORT DATE July 2010	3. REPORT TYPE AND DATES COVERED Draft Report October 2008 – August 2009	
4. TITLE AND SUBTITLE Low-Cost Warning Device Industry Assessment		5. FUNDING NUMBERS RR97A1/FG276	
6. AUTHOR(S) Adrian D. Hellman and Tashi Ngamdung			
7. PERFORMING ORGANIZATION NAME(S) AND ADDRESS(ES) U.S. Department of Transportation Research and Innovative Technology Administration John A. Volpe National Transportation Systems Center Cambridge, MA 02142		8. PERFORMING ORGANIZATION REPORT NUMBER DOT-VNTSC-FRA-09-12	
9. SPONSORING/MONITORING AGENCY NAME(S) AND ADDRESS(ES) U.S. Department of Transportation Federal Railroad Administration Office of Research and Development 1200 New Jersey Avenue, SE Washington, DC 20590		10. SPONSORING/MONITORING AGENCY REPORT NUMBER DOT/FRA/ORD-10/06	
11. SUPPLEMENTARY NOTES Safety of Highway-Railroad Grade Crossings series Program Manager: Leonard Allen			
12a. DISTRIBUTION/AVAILABILITY STATEMENT This document is available to the public through the FRA Web site at http://www.fra.dot.gov.		12b. DISTRIBUTION CODE	
13. ABSTRACT (Maximum 200 words) Under direction of the Federal Railroad Administration's Office of Research and Development, the U.S. Department of Transportation's Research and Innovative Technology Administration's John A. Volpe National Transportation Systems Center conducted a technology assessment of low-cost active warning devices for application at passive highway-rail grade crossings. The objective of this research was to present an objective assessment of the available low-cost warning device technologies and recommend a migration path that would facilitate implementation in the United States.			
14. SUBJECT TERMS Highway-rail, grade crossing, low-cost warning devices, railroad safety		15. NUMBER OF PAGES 47	
		16. PRICE CODE	
17. SECURITY CLASSIFICATION OF REPORT Unclassified	18. SECURITY CLASSIFICATION OF THIS PAGE Unclassified	19. SECURITY CLASSIFICATION OF ABSTRACT Unclassified	20. LIMITATION OF ABSTRACT

NSN 7540-01-280-5500

Standard Form 298 (Rev. 2-89)
Prescribed by ANSI Std. 239-18
298-102

METRIC/ENGLISH CONVERSION FACTORS

ENGLISH TO METRIC

LENGTH (APPROXIMATE)
- 1 inch (in) = 2.5 centimeters (cm)
- 1 foot (ft) = 30 centimeters (cm)
- 1 yard (yd) = 0.9 meter (m)
- 1 mile (mi) = 1.6 kilometers (km)

AREA (APPROXIMATE)
- 1 square inch (sq in, in^2) = 6.5 square centimeters (cm^2)
- 1 square foot (sq ft, ft^2) = 0.09 square meter (m^2)
- 1 square yard (sq yd, yd^2) = 0.8 square meter (m^2)
- 1 square mile (sq mi, mi^2) = 2.6 square kilometers (km^2)
- 1 acre = 0.4 hectare (he) = 4,000 square meters (m^2)

MASS - WEIGHT (APPROXIMATE)
- 1 ounce (oz) = 28 grams (gm)
- 1 pound (lb) = 0.45 kilogram (kg)
- 1 short ton = 2,000 pounds (lb) = 0.9 tonne (t)

VOLUME (APPROXIMATE)
- 1 teaspoon (tsp) = 5 milliliters (ml)
- 1 tablespoon (tbsp) = 15 milliliters (ml)
- 1 fluid ounce (fl oz) = 30 milliliters (ml)
- 1 cup (c) = 0.24 liter (l)
- 1 pint (pt) = 0.47 liter (l)
- 1 quart (qt) = 0.96 liter (l)
- 1 gallon (gal) = 3.8 liters (l)
- 1 cubic foot (cu ft, ft^3) = 0.03 cubic meter (m^3)
- 1 cubic yard (cu yd, yd^3) = 0.76 cubic meter (m^3)

TEMPERATURE (EXACT)
$[(x-32)(5/9)]\,°F = y\,°C$

METRIC TO ENGLISH

LENGTH (APPROXIMATE)
- 1 millimeter (mm) = 0.04 inch (in)
- 1 centimeter (cm) = 0.4 inch (in)
- 1 meter (m) = 3.3 feet (ft)
- 1 meter (m) = 1.1 yards (yd)
- 1 kilometer (km) = 0.6 mile (mi)

AREA (APPROXIMATE)
- 1 square centimeter (cm^2) = 0.16 square inch (sq in, in^2)
- 1 square meter (m^2) = 1.2 square yards (sq yd, yd^2)
- 1 square kilometer (km^2) = 0.4 square mile (sq mi, mi^2)
- 10,000 square meters (m^2) = 1 hectare (ha) = 2.5 acres

MASS - WEIGHT (APPROXIMATE)
- 1 gram (gm) = 0.036 ounce (oz)
- 1 kilogram (kg) = 2.2 pounds (lb)
- 1 tonne (t) = 1,000 kilograms (kg) = 1.1 short tons

VOLUME (APPROXIMATE)
- 1 milliliter (ml) = 0.03 fluid ounce (fl oz)
- 1 liter (l) = 2.1 pints (pt)
- 1 liter (l) = 1.06 quarts (qt)
- 1 liter (l) = 0.26 gallon (gal)
- 1 cubic meter (m^3) = 36 cubic feet (cu ft, ft^3)
- 1 cubic meter (m^3) = 1.3 cubic yards (cu yd, yd^3)

TEMPERATURE (EXACT)
$[(9/5)\,y + 32]\,°C = x\,°F$

QUICK INCH - CENTIMETER LENGTH CONVERSION

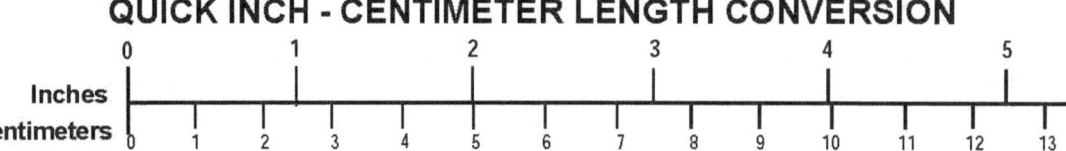

QUICK FAHRENHEIT - CELSIUS TEMPERATURE CONVERSION

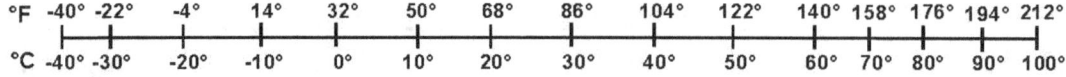

For more exact and or other conversion factors, see NIST Miscellaneous Publication 286, Units of Weights and Measures. Price $2.50 SD Catalog No. C13 10286 Updated 6/17/98

Acknowledgments

The U.S. Department of Transportation (USDOT) Federal Railroad Administration (FRA) Office of Research and Development sponsored the work leading to this report. The authors would like to thank Sam Alibrahim, Chief of the Signals, Train Control, and Communications Division, FRA, and Leonard Allen, Program Manager, Signals, Train Control, and Communications Division, FRA, for their guidance and direction in developing this report.

The authors also wish to thank Daniel Lafontaine, Chief Engineering, Grade Crossing and Access Control Programs, Transport Canada (TC), and Daniel Blais of the TC Transport Development Centre for their cooperation and contribution of data that was used for this project.

The authors would also like to acknowledge Marco daSilva, Highway-Rail Grade Crossing and Trespass Safety Research Program Manager, Systems Engineering and Safety Division, USDOT Research and Innovative Technology Administration's John A. Volpe National Transportation Systems (Volpe Center), Anya A. Carroll, Principal Investigator of Surface Transportation Programs, Center of Innovation for Physical Infrastructure Systems, Volpe Center, John McGuiggin, Chief of the Systems Engineering and Safety Division, Volpe Center, and Glenn Goulet, Chief of the Infrastructure and Facility Engineering Division, Volpe Center for their leadership and direction.

Contents

Executive Summary 1

1 Introduction 2

2 Discussion of Minimum Requirements and Cost 5

3 Analysis of Previous Research 8
   3.1 Federal Experience 8
   3.2 Transporation Review Board 10
      3.2.1 High-Speed Rail Innovations Deserving Exploratory Analysis Program 10
      3.2.2 National Cooperative Highway Research Program 11
   3.3 States 13
      3.3.1 Minnesota Experience 13
      3.3.2 North Carolina 16
   3.4 Transport Canada Research 18
   3.5 Australia 20

4 Current Research 22

5 Impediments to Acceptance 24
   5.1 Institutional 24
   5.2 Cost 25
   5.3 Technological 25

6 Findings 28

7 Recommendations 30

References 33

Appendix A. Sample Performance Criteria 36

Abbreviations and Acronyms 39

Figures

Figure 1. Twenty-Year Incident Rate at Active and Passive Public Grade Crossings (Ngamdung, 2009b) 3

Figure 2. Distribution of Grade Crossing Subsystem and Labor Costs (Petit, 2002) 7

Figure 3. Detail of Road Crossing Showing Island Limits and Vehicle Detection Limits for TTC Testing (Reiff, Gage, Caroll, and Gordon, 2003) 9

Figure 4. Schematic Representation of Double-Wheel Sensor Axle Counter Test Layout (Reiff, Gage, Caroll, and Gordon, 2003) 10

Figure 5. Mn/DOT Grade Crossing Warning Lights (left) and Advance Warning Sign Lights (right) (URS and TranSmart, 2005) 14

Figure 6. EVA Signal System Corporation Warning Device Configuration as Tested in North Carolina (Jennings, Field, Worley, and Scott, 2005) 17

Figure 7. TC Low-Cost Warning Device Configuration (Hildebrand, Roberts, and Robichaud, 2007) 19

Tables

Table 1. FRA Grade Crossing Signal System Safety Regulations 6

Table 2. Technologies Evaluated in TTI Research (Roop, Roco, Olson, and Zimmer, 2005)... 12

Table 3. High-Level Performance Criteria Categories Defined by TTI (Roop, Roco, Olson, and Zimmer, 2005) 12

Executive Summary

Virtually all of the grade crossing train detection and warning systems in the United States use a variant of the track-circuit technology developed over a century ago. Track circuits have evolved through the years, but the design and principles of operation have changed little. Although highly reliable, these systems are costly to install at low-usage grade crossings. Systems that leverage cost-effective, nontraditional technologies are an attractive alternative. The challenge for the railroad industry is to develop systems that are economical and safe.

In response to this challenge, the Federal Railroad Administration's (FRA) Office of Research and Development directed the U.S. Department of Transportation's (USDOT) Research and Innovative Technology Administration's John A. Volpe National Transportation Systems Center (Volpe Center) to conduct a technology assessment of low-cost active warning devices for application at passive highway-rail grade crossings. The objective of this research was to present an objective assessment of these technologies and recommend a course of action that would facilitate implementation in the United States.

Although there is no "low-cost" threshold, the body of research has shown a range of 5–30 percent of the cost for a conventional track-circuit–based grade crossing system. The large deviation was attributed to the variation in performance and functional requirements at each application. One of these is the location of the train detection and warning equipment, which has been developed for both on the right-of-way (ROW) or off-ROW environments. On-ROW systems are installed on railroad property and typically interface physically or electrically, with railroad infrastructure. Off-ROW systems are located external to railroad property and provide nonobtrusive train detection and warning functionality.

Many innovative on-ROW and off-ROW prototype systems have undergone extensive testing in North America, Europe, and Australia. However, a variety of technical, economic, and institutional issues need to be overcome before these technologies are considered mature enough to be adopted by railroads and government regulatory agencies. In recent years, regulatory bodies have become increasingly sophisticated in their knowledge of nonconventional train detection and warning technologies. This is reflected in the growing use of performance-based regulations, which offer more flexibility for railroads and railroad suppliers to demonstrate safety.

1. Introduction

Passive highway-rail grade crossings account for approximately 50 percent of the total number of public crossings in the Unites States, according to USDOT's FRA (2009a). Over the past 20 years, Federal, State, and local governments, as well as railroads, have succeeded in significantly reducing accident risk posed by passive grade crossings. The most successful strategy has been to close crossings wherever possible and to install active warning devices at locations that cannot be closed under any circumstances. This is borne out in statistics maintained by FRA.

Of the 178,000 public grade crossings in the United States in 1989, 120,000 or 67 percent were designated as passive in the FRA highway-rail grade crossing inventory. By 2008, this number was reduced by 40 percent to 71,000 or 52 percent of the 137,000 public grade crossings. Coincidentally, the number of active public crossings increased from 58,000 to 66,000 over the same 20-year period (Ngamdung, 2009a). Other strategies, including innovative placement of stop and yield signs, have proven effective in reducing risk as well (Russell, Rys, and Liu, 1999).

The plots in Figure 1 signify the 20-year decline in incident rates at passive and active grade crossings, where the incident rate data have been normalized with respect to traffic moment (T-M). Incidents at active grade crossings, denoted by the red line, decreased from approximately 9 per 100 T-M to 2 per 100 T-M, roughly equivalent to an 80 percent reduction. Meanwhile, incidents at passive crossings decreased in a similar fashion from 85 per 100 T-M to 15 T-M, resulting in an 80 percent drop as well (Ngamdung, 2009b). Despite these successes, the incident rate for passive grade crossings still exceeds that of active ones by a factor of 10, and the magnitude ratio has not varied substantially.

The decrease in the passive crossing incident rate reflects the concerted effort to either eliminate the riskiest crossings or upgrade them with active warning devices. The remaining group of passive crossings poses less of a risk but still requires serious attention. Working against this trend is the limited availability of Federal and State funding for crossing improvements as well as the competition with other highway improvement programs for funding. This has necessitated that grade crossings be prioritized for improvement in terms of risk and the funds allocated accordingly, resulting in slow but steady progress. This is the environment that forms the basis for low-cost warning device research.

Virtually all of the grade crossing train detection and warning systems in the United States employ a variant of the same basic track-circuit technology developed a century ago. In recent decades, railroads have migrated from relay-based control systems to those that are processor-based. However, the design and principles of operation have changed little. The procurement and installation cost of these traditional crossing systems is the primary impediment to more widespread deployment. According to a 1995 U.S. General Accounting Office report, the average cost of installing flashing lights and gates was $150,000 (1995). If a modest 3 percent inflation rate is assumed, the equivalent cost in current dollars is $230,000. This is an average value and can vary greatly depending on the complexity of the circuit logic, the type of warning device installed (flashing lights, gates, etc.), and the labor effort required. However, assuming that this is somewhat reflective of actual values, the cost to install flashing lights and gates at the 71,000 existing passive crossings would be approximately $16 billion. This is four times greater than the estimated $4 billion Federal and State investment in grade crossing improvements since the Rail-Highway Crossing Program began in 1974 (FRA, 2009b). At these pricing levels, there is little benefit-cost justification for upgrading many passive grade crossings with active warning devices.

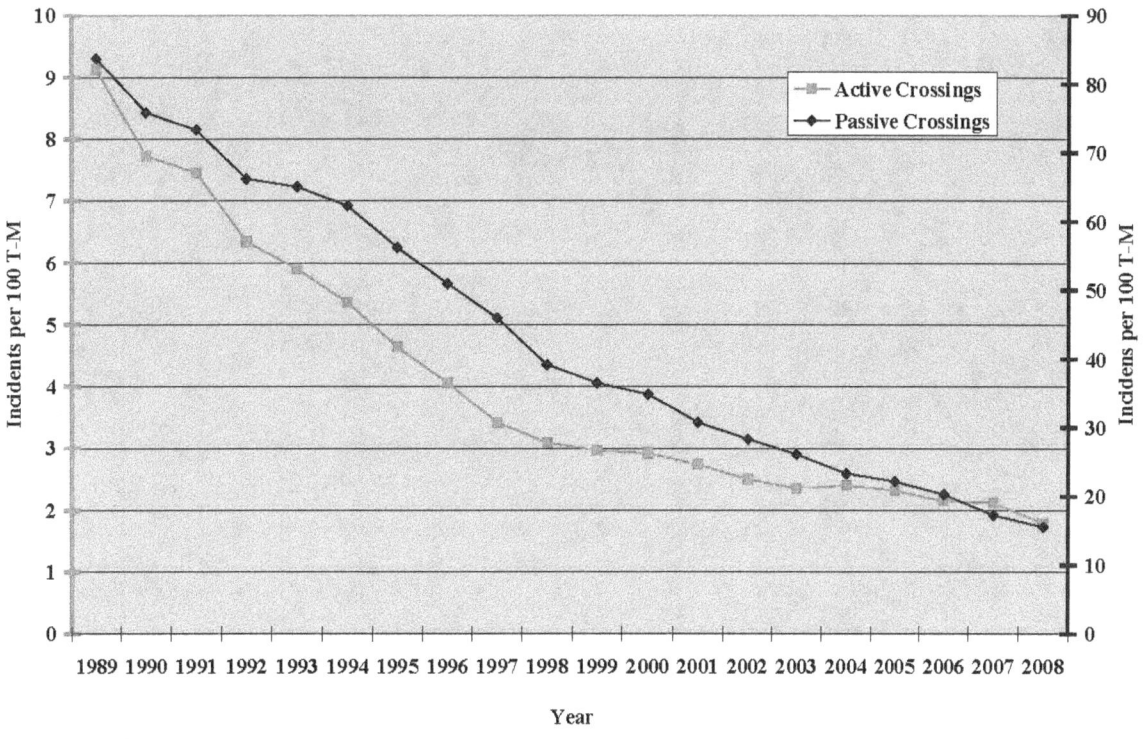

Figure 1. Twenty-Year Incident Rate at Active and Passive Public Grade Crossings (Ngamdung, 2009b)

This environment has fostered a great deal of interest and research in the viability of low-cost risk mitigation alternatives at passive grade crossings, many of which are located in rural areas of the country. A late 1990s National Transportation Safety Board (1998) report included among its recommendations the completion of standards development for applications of intelligent transportation systems (ITSs) at grade crossings and the installation of stop and stop ahead signs at all passive crossings. ITS solutions usually consist of an in-vehicle warning system that communicates directly with an approaching train. This architecture circumvents the need for traditional train detection and warning systems. Consequently, the ITS was considered to present a low-cost opportunity to provide some type of active warning at passive crossings. However, the requirement that all trains and motor vehicles be equipped has proven to be a difficult obstacle to overcome. Considerable research has been performed to characterize the safety benefits of supplementing the basic cross buck with stop and stop ahead signs. The efficacy of these signs has been debated for many years (Lerner, Llaneras, McGee, and Stephens, 2002). However, recent research has shown that stop sign-controlled crossings provide a statistically significant 45 percent reduction in accident rate over cross buck-only controlled crossings (Millegan, Yan, Richards, and Han, 2009). This is a positive trend but still leaves the passive crossing accident rate at a much higher baseline than active crossings.

2. Discussion of Minimum Requirements and Cost

One of the most compelling discussions of low-cost warning devices was held at a forum associated with the 2001 National Highway-Rail Grade Crossing Safety Conference. This forum brought together an assortment of experts from railroads, suppliers, and researchers. Bill Peterson of the Burlington Northern Santa Fe (BNSF) railroad presented some unique insights into the railroad perspective. One of his key points related to the difference in motorist perception to grade crossing warning systems and highway traffic lights. When highway traffic lights at an intersection are dark, motorists know that they are required to reach a full stop and proceed only when the intersection is free of traffic. However, a motorist approaching an active grade crossing with darkened warning devices perceives this situation as safe to proceed without stopping, whether or not the equipment is functioning properly.

The uniqueness of the grade crossing environment underscores the need to ensure that the warning devices are highly reliable and fail-safe. These requirements are expensive to satisfy and represent some of the primary roadblocks to the implementation of low-cost warning systems. Most other requirements, except for the cost associated with installation, dovetail from these two. For example, new active grade crossing systems require a new connection to the electrical power grid, typically at a cost of approximately $10,000. Although solar and wind power have been the subject of interest in recent years, the reliability of these systems, even with a battery or generator backup, is insufficient (Peterson, 2001).

As presented by Moody and Reiff (2001) at the same forum, track-circuit–based grade crossing systems range in cost from $100,000 to $150,000. They suggested a range of $20,000 to $50,000 for low-cost warning devices with reduced features that would satisfy a minimum set of performance requirements as governed by the FRA regulation for grade crossing safety. These requirements include a minimum warning time of 20 seconds, incorporation of fail-safe operation principles, noninterference with existing systems and signals, and use of applicable design requirements found in the *Manual of Uniform Traffic Control Devices* (MUTCD), Parts 8 and 10.

Annual maintenance costs, which are usually financed by the railroad, may be even more critical than initial system cost. For low-cost warning systems, these must be significantly less than for traditional grade crossing systems for the systems to still be considered "low cost" (Moody and Reiff, 2001).

In 2002, the American Railway Engineering and Maintenance-of-Way Association (AREMA) Committee 36, Highway-Rail Grade Crossing Warning Systems, published an analysis of baseline grade crossing system costs for evaluating the costs of current and future technologies. The objective of the analysis was to determine the relative cost of grade crossing subsystems and the related costs of design, engineering, and installation services (Petit, 2002).

The first step of the analysis entailed identifying the minimum requirements, in terms of regulatory compliance and industry guidelines, that all grade crossing systems must satisfy. From the regulatory perspective, Title 49, Part 234 of the Code of Federal Regulations (CFR), Grade Crossing Signal System Safety, "imposes minimum maintenance, inspection, and testing standards for highway-rail grade crossing warning systems." The maintenance, inspection, and testing standards are found in Part 234, Subpart D. The sections relevant to minimum requirements are shown in Table 1 below.

Table 1. FRA Grade Crossing Signal System Safety Regulations

Section	Description
234.203	All control circuits that affect the safe operation of a highway-rail grade crossing warning system shall operate on the fail-safe principle.
234.205	Operating characteristics of electromagnetic, electronic, or electrical apparatus of each highway-rail crossing warning system shall be maintained in accordance with the limits within which the system is designed to operate.
234.215	A standby source of power shall be provided with sufficient capacity to operate the warning system for a reasonable length of time during a period of primary power interruption.
234.225	A highway-rail grade crossing warning system shall be maintained to activate in accordance with the design of the warning system, but in no event shall it provide less than 20 seconds warning time for the normal operation of through trains before the grade crossing is occupied by rail traffic.
234.227 (a)	Train detection apparatus shall be maintained to detect a train or rail car in any part of a train detection circuit, in accordance with the design of the warning system.
234.275	This requires that grade crossing warning systems, subsystems, or components that are processor based, which contain new or novel technology, are required to comply with the safety and risk analysis requirements defined in 236, Subpart H. In this context, new or novel technology specifically refers to designs that do not use conventional track-circuit technology.

These standards are quite rigorous and must be satisfied by any warning system, regardless of the cost. In its report, AREMA Committee 36 addressed all of these requirements, as well as industry best practices, to evaluate grade crossing system cost. The minimum requirements described above were used to develop a baseline cost allocation with the following attributes:

- Single-track application with dual direction operation possible.
- Warning devices consisting of entrance gates, flashing lights, and bells.
- Train detection system with conventional track-circuit technology for uniform time warning. The approach circuit is approximately 3,000 feet in each direction, with an

island circuit overlay. Both nonredundant and redundant train detection systems were considered.
- External event recorder.
- Processor or relay-based independent controller for warning devices.
- Full battery backup with 48 hours of capacity.
- Equipment installed in a 6 × 6-foot bungalow.
- Crossing not located close to another crossing, requiring communication between the two.

The relative cost distribution for nonredundant grade crossing systems is shown in Figure 2. As described by Petit (2002), installation, engineering, freight, and power service account for at least 50 percent of the cost of a grade crossing. Installation cost is typically considered fixed, because many railroads are bound by labor agreements to use railroad signal installers rather than outside contractors. This can range from roughly 25 to 35 percent of the total cost of a grade crossing system. In contrast, the cost of grade crossing electronics, including the train detection, crossing controller, and event recorder subsystems, varies from 12.5 to 17 percent of the total system cost, depending on the level of redundancy required in the detection system. Train detection, which is frequently targeted for cost reduction in low-cost warning system research, accounts for no more than 11.5 percent of system cost.

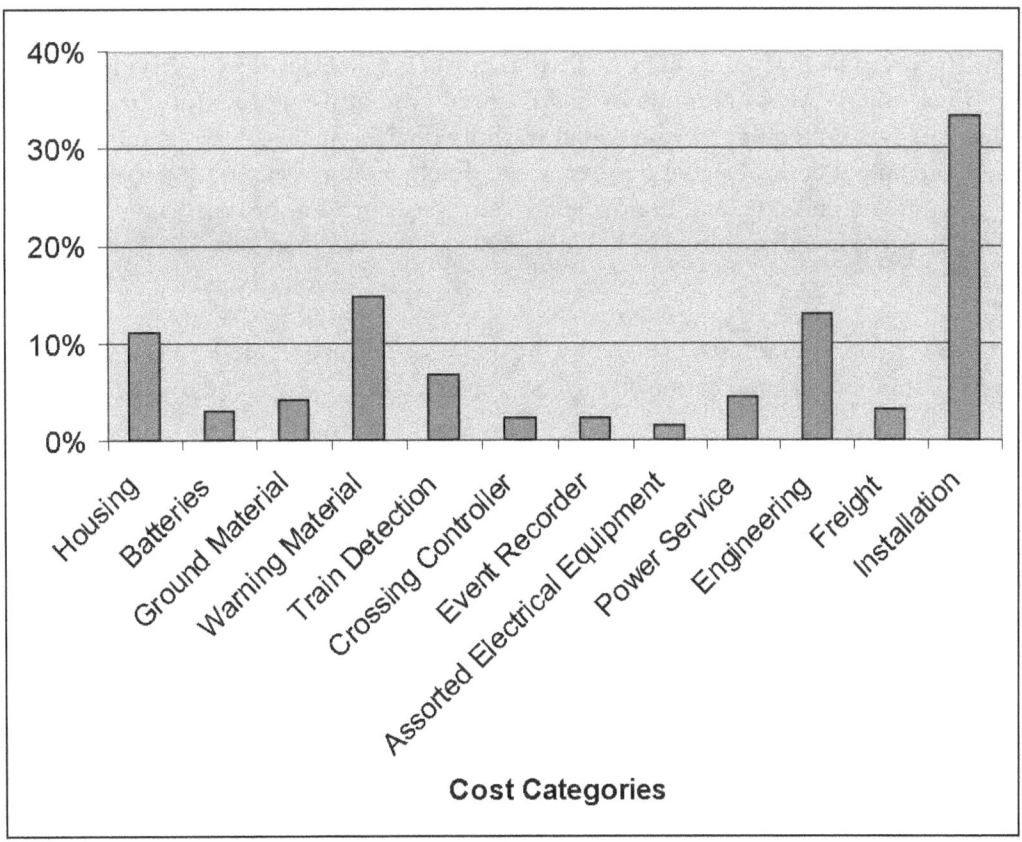

Figure 2. Distribution of Grade Crossing Subsystem and Labor Costs (Petit, 2002)

3. Analysis of Previous Research

3.1 Federal Experience

The requirements presented by Moody and Reiff at the 2001 National Highway-Rail Grade Crossing Safety Conference were developed as part of a larger research program at FRA's Transportation Technology Center (TTC) in Pueblo, CO. In this FRA research initiated in 1999, the Volpe Center and the Transportation Technology Center, Inc. (TTCI), evaluated the performance of non-track-circuit–based train and vehicle detection technologies (Reiff, Gage, Carroll, and Gordon, 2003).

In February and March 1999, TTCI issued a Request for Technical Information (RFTI) to almost 300 prospective vendors who had participated in previous grade crossing testing. The RFTI was specifically written to solicit information for train detection alternatives to track circuits and identified the salient minimum performance and operating requirements for which products would be evaluated. These included:

- Minimum train approach warning time of 20 seconds
- Release of island detection within 2 seconds of train departure
- Train detection speed regimen of 5–125 miles per hour (mph) (8–200 kilometers per hour (km/h))

Of the 20 proposals that were submitted, six finalists were selected by the evaluation team. Each company was required to install a prototype of its product for testing at TTC. The test site was not equipped with actual warning devices such as gates and lights. Instead, the output control signal was recorded. Each vendor was required to submit a reliability, maintainability, and life-cycle cost analysis. In addition, each vendor was required to provide a description of how its product was designed using fail-safe principles.

Only five vendors submitted a product by the deadline defined in the RFTI, and only four included a train detection component. These consisted of:

- Two systems using vibration and magnetic anomaly sensors
- One using inductive loops
- One using double-wheel sensors (axle counters) for train detection

The test environment is shown in Figure 3.

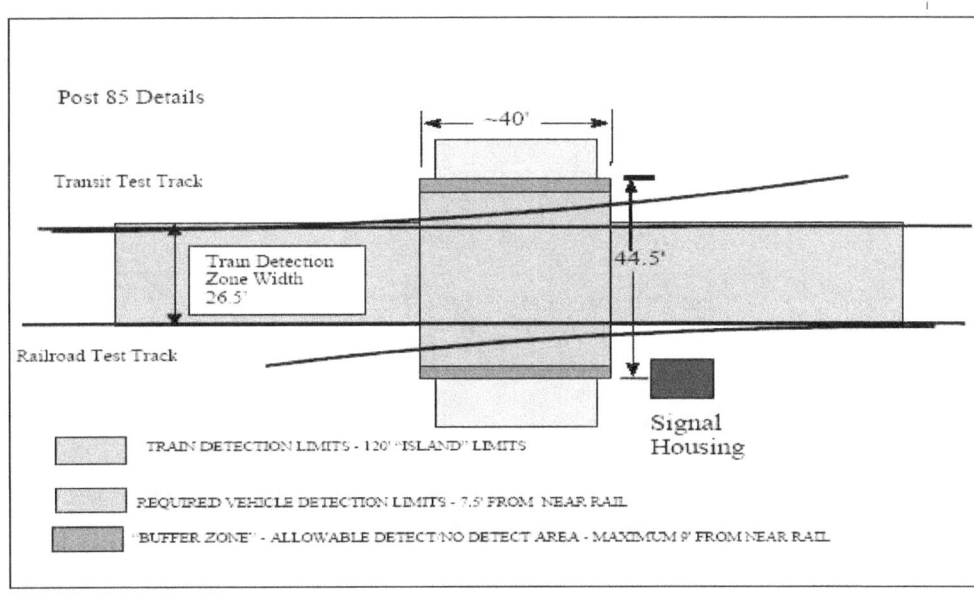

Figure 3. Detail of Road Crossing Showing Island Limits and Vehicle Detection Limits for TTC Testing (Reiff, Gage, Caroll, and Gordon, 2003)

All three systems were designed to provide fail-safe functionality, as described by each respective vendor. In addition to the minimum performance requirements listed above, each system was evaluated in terms of successful detections, critical failures, missed detections, and nuisance or false alarms. These measurements were performed for both approach and island indication detection. The double-wheel sensor was the only system able to detect all of the approach and island activations with no failures, missed detections and false alarms. The test configuration, shown in Figure 4, consisted of six total double-wheel sensors, a redundant pair at each approach, and one on either side of the island approximately 20 feet from the crossing. In simple terms, the approach sensors were used for train detection and warning device activation, with the redundant sensors serving as backups in case the first ones were not working properly. As the train passed over the grade crossing approach, the sensors counted the number of axles and transmitted the result to the grade crossing controller. This process was repeated as the train passed over each of the island sensors. Once the last car had exited the crossing, the controller compared the approach and island axle counts. If the values were equal, the controller released the crossing and deactivated the warning devices.

This technology is widely used in Western European countries in overlay (France) and stand-alone (Germany and Spain) configurations (Luedeke, Wagner, Carroll, and Markos, unpublished manuscript). However, it has been received less favorably in the United States, mostly because of reliability concerns such as sensor spacing relative to wheel flanges, debris buildup on the magnetic components of the sensors, and damage resulting from ice, rocks, and dragging of equipment. In recent years, design improvements have shown promise in increasing wheel sensor reliability. Another, possibly more significant concern, is that railroads in the United States view the continuous detection characteristic of track circuits more favorably than the point detection of wheel sensors (Roop, Roco, Olson, and Zimmer, 2005). This is particularly critical in the context of designing train detection systems that satisfy FRA grade crossing signal system safety regulations.

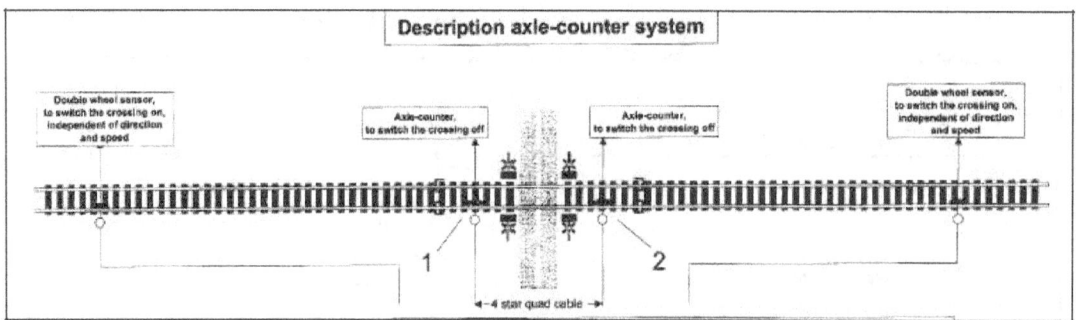

Figure 4. Schematic Representation of Double-Wheel Sensor Axle Counter Test Layout (Reiff, Gage, Caroll, and Gordon, 2003)

The authors of the report found many of the technologies to be promising. However, because this was a research program, they did not endorse any of them nor suggest how they may be implemented by the railroad industry (Reiff, Gage, Caroll, and Gordon, 2003). This research was preceded by testing of island detection system alternatives to track circuits performed by the Association of American Railroads (AAR) in 1996. Four technologies were tested:

- Count-in/count-out using strain gages
- Count-in/count using magnetic wheel sensors
- Modification of the existing island circuit
- Combination of movement and infrared presence detection

During preliminary testing, the infrared system was found to be inadequate in train presence detection reliability and was removed from the test program. The three remaining technologies were then tested at an active grade crossing test site on the CSX, BNSF, and Union Pacific railroads, respectively. All three systems were installed at each test site in parallel with the conventional grade crossing circuit. The results of this research showed that none of these technologies could reliably detect trains as well as track circuits. As a final note, the authors recommended that these technologies be monitored for improvements in reliability (Guins, Reiff, Gurule, and Gage, 1996).

3.2 Transportation Review Board

3.2.1 High-Speed Rail Innovations Deserving Exploratory Analysis Program

FRA has used its affiliation with the Transportation Research Board (TRB) to sponsor low-cost warning device research as part of a greater focus on enabling railroad technologies. One such mechanism was the Innovations Deserving Exploratory Analysis (IDEA) program, which targeted promising but unproven concepts with the potential to advance surface transportation systems. In 1997, FRA began funding the High-Speed Rail IDEA program in which research proposals are selected on the basis of their potential role in upgrading the existing U.S. rail system to accommodate operations up to 100 mph (160 km/h) and beyond. Although low-cost grade crossing detection and warning systems were not a primary objective of this program, several concepts with potential application toward train detection were funded and completed. The technologies studied in this research included:

- Buried fiber-optic cable
- Rail-mounted fiber-optic cable
- Microwave radar

The objective of the buried fiber-optic cable research, performed by the Texas Transportation Institute (TTI) and completed in 2001, was to determine its potential to provide an inexpensive, reliable alternative to conventional track circuitry for train presence and broken rail detection. However, an excess of environmental noise so severely constrained the detection range that the research was concluded after laboratory testing (Olson and Roop, 2003). Concurrently, the University of Illinois investigated the viability of installing fiber-optic cable directly to rail. This research, completed in 2001, was more successful in minimizing environmental noise than the buried cable project. However, the differences were not readily apparent and a host of other issues including cost and protecting the cable from damage remained (Chuang and Young, 2003).

From 1997 to 2000, O'Conner Engineering, Inc., developed and tested a microwave Doppler radar system for train detection. The radar specification included a range of about 1 mile with a 2-foot resolution and the capability to detect approaching or departing trains at velocities up to 150 mph (240 km/h). O'Conner Engineering performed testing at a single-track and double-track grade crossing. However, the test suite did not include characterizing the ability of the radar technology to discriminate between two or more trains within the field of view. The results showed that the radar suffered from detection limitations specific to high-speed trains and grade crossings with curved or obstructed approaches (O'Conner, 2009). This technology was documented in a research report of low-cost warning devices published by TTI (Roop, Roco, Olson, and Zimmer, 2005).

3.2.2 National Cooperative Highway Research Program

In 2004, TTI was funded by the TRB National Cooperative Highway Research Program to identify and assess low-cost, viable, active, warning system and component designs for highway-rail grade crossings. The first report stemming from this research was published in 2005. In the report, the TTI researchers evaluated 12 potential low-cost grade crossing warning device technologies against a set of cost and performance criteria they had developed. These technologies are listed in Table 2 on the next page, and the general performance criteria categories are shown in Table 3. A more detailed description of the sample performance criteria is provided in the appendix to this report.

Table 2. Technologies Evaluated in TTI Research
(Roop, Roco, Olson, and Zimmer, 2005)

Technology	On-ROW	Off-ROW
Geophone	√	
Fiber-optic (rail)	√	
Fiber-optic (buried)	√	
Video imagery		√
Radar (speed)		√
Radar (speed and distance)		√
Acoustic		√
Pressure sensor	√	
Magnetic anomaly	√	
Infrared		√
Laser (lidar)		√
Ultrasonic		√

Table 3. High-Level Performance Criteria Categories Defined by TTI
(Roop, Roco, Olson, and Zimmer, 2005)

Objectives	Criteria
Enhanced Safety	Fail-safe design
	Provides 20-second advanced warning
	Geometric flexibility
Reduced System Cost	Installation cost
	System life
Reliability	Failure rate
	Redundancy
	Simplicity
	Disruption resistance
	Known and controlled-for failure modes
Installability	Ease of installation
	Site preparation
Maintainability	Ease of inspection and testing
	Ease of repair
Compatibility	Encroaches on railroad property

In contrast to the AREMA Committee 36 research, TTI defined the baseline system as train detection and flashing lights and/or warning lights, but no gates. For this platform, TTI established several cost categories: (1) ultralow, less than $25,000, (2) low, $25,000 to $50,000, and (3) moderate, $50,000 to $75,000 (Roop, Roco, Olson, and Zimmer, 2005). This compares favorably with values suggested by Reiff and Moody (2001).

The results of this research were divided between systems that were installed on the railroad (ROW) and off-ROW. The railroad industry aversion to risk, resulting from tort liability concerns, suggested that off-ROW systems may be easier to implement. However, the dichotomy between the on-ROW and off-ROW systems, especially in regards to safety, was significant.

In terms of safety, the best performing ultralow-cost technologies were fiber-optic (rail) and fiber-optic (buried), both of which must be installed on railroad property. For the low-cost category, the pressure sensor and magnetic anomaly technologies received the highest safety rating. Both of these technologies are required to be installed on the ROW as well. Only in the moderate cost category were two off-ROW technologies, ultrasonic and radar, found to rank highest overall. The off-ROW technologies, specifically the acoustic, ultrasonic, and radar, fared considerably better in the reliability rankings.

In the next phase of this research, conducted in 2006, TTI tested the radar and acoustic detection systems at a four-quadrant gate crossing in College Station, TX. These technologies were selected because they could be installed off-ROW and not be constrained by railroad liability issues. The equipment for both systems was mounted on a city traffic light utility pole at a height of 14 feet above the ground and 20 feet from the nearest rail. For the radar system, an antenna was installed on the utility pole to detect an approaching train in either direction, and a single antenna was installed for island detection. The radar technology was capable of train detection and speed measurement. The acoustic system consisted of a single omnidirectional microphone for detection of locomotive train horns and was only capable of train detection.

The performance of each of these technologies was compared with the track-circuit signaling system that was already installed at the grade crossing and was recorded in shadow mode. The data collection period, 76 days between August and October 2006, consisted of approximately 1,500 track-circuit activations by approaching trains. Both systems detected 100 percent of the trains but were much less successful at discerning rail from highway traffic. This was marked by the high false train detection rates shown by both systems, 57 percent for the radar system and 94 percent for the acoustic system. Although the radar system demonstrated more potential, neither was considered robust enough to satisfy railroad liability concerns in spite of being installed off-ROW.

3.3 States

3.3.1 Minnesota Experience

From 2001 to 2005, the Minnesota Department of Transportation (Mn/DOT) developed and deployed a low-cost warning system on the Twin Cities and Western Railroad (TC&W). The objective of this program was to compare the performance of this technology with that of traditional track-circuit–based warning devices at grade crossings with few train movements and highway users. Mn/DOT defined "low cost" as $10,000 to $15,000, or the equivalent to 10 percent the cost of a conventional grade crossing warning system (URS and TranSmart, 2005).

In contrast to the wayside-centric functionality of most train detection and warning systems, this system was locomotive-centric. One of the core features of this type of system was that each equipped locomotive resolved its location as well as the location of all grade crossings. This was accomplished by means of a global positioning system (GPS) receiver installed on every locomotive. Another core feature was the use of wireless radio technology for communications between locomotives and grade crossings.

Although gates were not installed at any of the grade crossings, they were all upgraded with the traditional cross buck-mounted flashing lights facing each direction of oncoming highway traffic, as shown in Figure 5. As an added safety measure, an active warning sign, connected to the warning system, was installed on each approach lane in advance of the crossings, also shown in Figure 5. All of the crossing equipment, including the controller, was powered by a 14-volt battery with a solar panel charging system (URS and TranSmart, 2005).

Figure 5. Mn/DOT Grade Crossing Warning Lights (left) and Advance Warning Sign Lights (right) (URS and TranSmart, 2005)

Unlike wayside-centric systems, train detection was performed by the onboard locomotive equipment. In this multistep process, the GPS receiver, in conjunction with a dead reckoning system, determined locomotive location, travel direction, and speed. The onboard equipment used these data to resolve the locomotive position with respect to nearby grade crossings and calculated an estimated time to arrival. This was performed dynamically and therefore provided for grade crossing warning device activation using a constant warning time (CWT) algorithm. Train detection was ultimately achieved by transmitting this information to the grade crossings. Although all grade crossings within 5 km of the locomotive receive the data, only crossings within 2 km were authorized to be activated. Because CWT was calculated on-the-fly, the grade crossing controller was programmed to activate the warning devices using a preset warning time as long as it meets the 20-second minimum requirement (URS and TranSmart, 2005).

Island detection was accomplished by an integrated magnetometer and ultrasonic sensor unit installed on each cross buck. The magnetometer was responsible for primary train detection in the island, whereas the ultrasonic sensor was used to confirm the magnetometer output. This subsystem was designed to operate independently of the train-borne detection and activation equipment and was not considered to be a primary crossing activation device (URS and TranSmart, 2005).

The train to grade crossing radio network was designed on the shared network model and operated in the 220-megahertz ITS dedicated frequency band. Because the network was shared, multiple locomotives and grade crossings were supported. The one-way transmission range of the radio equipment was 5 km and allowed for locomotive and grade crossing equipment to establish a data exchange up to that limit. However, only crossings within a 2 km approach could be activated (URS and TranSmart, 2005).

If there was a system error involving the function of the locomotive or the grade crossing equipment, or difficulty establishing data communications between a locomotive and the crossing equipment, the crew was notified by a display and audible warning. The crew would then be required to stop the train on the crossing approach and manually flag the crossing (URS and TranSmart, 2005).

In total, 27 passive grade crossings on a 75-mile corridor of the TC&W between Cologne and Renville, MN, were equipped with the low-cost warning device system. Between April 2003 and November 2004, Mn/DOT incrementally installed and tested the grade crossing equipment, equipping 10 crossings first. After extensive shadow mode testing, the final 17 crossings were equipped and programmed to operate in shadow mode.

The field operational test (FOT) was performed during an 80-day span from June to September 2005 under a waiver from FRA. Six of the 27 crossings were operated in active mode. The remaining 21 crossings were operated in shadow mode and thus transparent to motorists. After the FOT was completed, the warning system was removed from all of the 27 crossings.

The results of this research, published in December 2005, indicated that no system failure was reported during the FOT. The active warning system accurately warned and provided adequate warning times to motorists for approaching trains. The research findings were submitted to FRA as part of the waiver and approval process (URS and TranSmart, 2005).

Although the overall highway rail interface HRI system offers great potential to a low-cost alternative, TC&W never demonstrated compliance with the FRA regulation for fail-safe grade crossings control circuitry, specifically train detection. In the initial waiver of compliance, TC&W sought regulatory relief from 49 CFR Part 234.203, which requires that all control circuits that affect the safe operation of a highway-rail grade crossing warning system shall operate on the fail-safe principle. In the waiver request, TC&W stated that the train detection system was not track-circuit based and thus not in strict compliance with the regulation. Rather, the railroad claimed that they complied with the "spirit and intent" of the regulation by means of a non-track-circuit–based train detection subsystem (FRA, 2002).

3.3.2 North Carolina

Between October 2001 and September 2003, the North Carolina Department of Transportation (NCDOT) tested a magnetometer-based train detection and warning system. Magnetometer technology is used to detect changes in the ambient magnetic field of the earth associated with the presence of a large mass iron, such as an approaching train. The technology, developed by EVA Signal System Corporation, was selected by NCDOT for several reasons. First, the EVA product offered a combination of lower cost and unique features not matched by the other systems evaluated by NCDOT. Second, the magnetometer technology was designed for off-track train detection and completely independent of the track circuitry. Finally, the technology was marketed as fail-safe by EVA Signal (Jennings, Field, Worley, and Scott, 2005).

The objectives of the test were to test the ability of the EVA system to detect trains reliably and accurately, test the ability of the systems to communicate effective warnings to motorists, and study the reaction and behavior of the public to the experimental signal. The equipment was installed at a grade crossing on the North Carolina and Virginia Railroad (NCVARR) near Rich Square, NC. From October 2001 to July 2003, the EVA 1000 system was tested (Jennings, Field, Worley, and Scott, 2005).

The warning system consisted of a traditional cross buck with flashing red light-emitting diodes (LEDs) in an "X" pattern, a flashing strobe light bar mounted 5 feet above the cross buck, and a train detection advisor mounted under the strobe bar, consisting of six sequentially activated yellow halogen lights to provide motorists with the direction of an approaching train. The whole grade crossing system was battery powered by three 12-volt batteries with an optional solar recharging system (Jennings, Field, Worley, and Scott, 2005). The configuration is shown in Figure 6.

The initial train detection configuration consisted of six magnetometer sensor probes buried in the railroad ROW adjacent to the NCVARR track. The probes were installed in the following arrangement on either side of the crossing:

- One at 1,690 feet (512 meters) and another at 1,590 feet (482 meters) from the crossings for calculation of train velocity
- One for verification of train movement at 1,000 feet (303 meters) from the crossing
- One for island activation at 50 feet (15 meters from the crossing)
- Two probes for backup detection at 1,540 feet (467 meters) and 150 feet (45 meters) from the crossing (Jennings, Field, Worley, and Scott, 2005)

Figure 6. EVA Signal System Corporation Warning Device Configuration as Tested in North Carolina (Jennings, Field, Worley, and Scott, 2005)

The sensor probes were hardwired by a buried cable to a control unit adjacent to the crossing, which was housed in a waterproof underground vault. The control unit employed the signals from the first two sensors to calculate train velocity and time to train arrival at the crossing. This provided CWT functionality and adequate warning times for the motorist. A pair of infrared sensors was installed at the crossing itself to verify train presence. These sensors were mounted to steel posts on either side of the crossing and were aimed diagonally across the tracks (Jennings, Field, Worley, and Scott, 2005).

From the outset of the testing, the EVA 1000 system was constantly plagued by operational issues, primarily from false activation of the warning devices and low island detection reliability. Although EVA Signal and NCDOT worked diligently, these issues could not be resolved. The high false activation rate was found to result from the inability of the magnetometer probe sensors to discriminate trains from maintenance-of-way (MOW) equipment and all-terrain vehicles (ATVs). EVA offered fencing as a possible solution for the ATV issue but does not seem to have ever been implemented by NCDOT or the railroad. In any case, this would not have resolved the false activations attributed to the MOW equipment. In July 2003, the crossing was upgraded with the EVA 3000 system that featured an improved island detection module. However, the test program was terminated before the effectiveness of the new island detection module could be evaluated. The low reliability of the technology precluded the third objective of the program, the driver behavior study, from ever being initiated (Jennings, Field, Worley, and Scott, 2005).

3.4 Transport Canada Research

From April 2005 to March 2007, the University of New Brunswick (Canada) evaluated a radar-based low-cost warning device system for private and farm crossings. This evaluation was unique to the others described in this report in that the test site was not an actual grade crossing, but rather a section of track. In this manner, the research team could measure the performance of the detection and warning device equipment without interfering with motor vehicle traffic. The actual test occurred in Grand Bay-Westfield, New Brunswick, on the New Brunswick Southern Railway, near St. John, New Brunswick. At the time of the testing, the line experienced a train frequency of 50–60 trains per month. With hi-rail vehicle traffic included, the total rail-borne movements could range from 60 to 90 trains per month (Hildebrand, Roberts, and Robichaud, 2007).

At the time the research was performed, passive private and farm road crossings were not under the regulatory control of Transport Canada (TC). However, a crossing that is upgraded with active warning devices falls within the jurisdiction of the TC grade crossing regulations. In a slight variation of the U.S. grade crossing safety regulations, TC regulations reference AAR (now AREMA) recommended practices for design and operational requirements for signals, gates, operating mechanisms, and control circuits. Likewise, the recommended practices require that these components be designed on the fail-safe principle (Hildebrand, Roberts, and Robichaud, 2007).

The system included train detection and warning device functionality. Like the other systems described thus far, there was a separate system for both train approach and island presence detection. A dual scanning X-band Doppler radar, one facing each direction of train travel, was used to detect both approaching and departing rail traffic. The radar operated in the 10.5-gigahertz band and had an adjustable detection range of 250–3,000 feet (76–909 meters). A single ultrasonic sensor, operating in the 50-kilohertz band was responsible for island presence detection. The warning devices consisted of electronic bells and flashing LED lights. The entire system was powered by two 120-ampere batteries supported by three 80-watt solar panels. All of the equipment, including the sensors, was mounted on a single pole on the ROW (Hildebrand, Roberts, and Robichaud, 2007).

The system, shown in Figure 7, was installed in April 2005 and field-tested through March 2007. Because no commercial low-cost warning device technology was available, the system tested was a prototype in shadow mode. The evaluation criteria consisted of assessing the system performance with respect to the train detection components, warning time adequacy, and solar panel power system effectiveness and reliability. Specific parameters that were captured included total number of system activations, duration of activations, missed detections, and clearance times (Hildebrand, Roberts, and Robichaud, 2007).

**Figure 7. TC Low-Cost Warning Device Configuration
(Hildebrand, Roberts, and Robichaud, 2007)**

High false train detection and missed activation rates were a constant issue throughout the entire system testing process, especially during initial testing. These were attributed to either the radar system detection software algorithm, which resulted in over- or undercalibration of the detection sensitivity, or faulty operation of the radars themselves. The software sensitivity issue was resolved by a series of four firmware updates, with the final implementing an adaptive detection algorithm. However, the reliability of the radar units remained a constant concern, with two failing during the 2-year evaluation period. Overall, the system showed some potential, but it is still not reliable and robust enough to be used as stand-alone active warning system in the United States. In addition, it does not incorporate fail-safe operation (Hildebrand, Roberts, and Robichaud, 2007).

A unique feature of this platform is that the manufacturer ships it preassembled, with the exception of the solar panel. This type of platform could potentially yield significantly reduced savings in installation related labor costs (Hildebrand, Roberts, and Robichaud, 2007).

3.5　Australia

In the early part of 1999, the State of Victoria started a trial of a low-cost warning device. The main objective was to develop a system that would cost about 1/10th of a conventional active system, be fail-safe, and most importantly, make the passive crossing conspicuous at the time a train is approaching the crossing. Five devices, consisting of a Doppler radar unit, two magnetometers, an in-train transmitter, and an electromagnetic induction loop, were screened based on reliability of train detection. The devices were installed on an abandoned section of track and subjected to approximately 500 passes by a hi-rail vehicle. The highest reliability technology, the electromagnetic induction loop, was selected for further testing (Jordan, 2006).

The system was tested over 1 week on an abandoned section of track using a road transferable locomotive. The test configuration consisted of a pair of inductive loop detectors installed between the rails and on top of the railroad ties 16.5 feet (5 meters) apart and approximately 2000 feet (606 meters) from the crossing. The detector pair added a measure of redundancy to the detection process but also allowed for calculation of train velocity.

A third detector installed at the crossing served a dual purpose: (1) turn off the flashing lights once a train has departed the crossing and (2) a redundant detection in case the first two detectors failed to detect a train. The signal from the detectors was transmitted to a grade crossing controller at the crossing by a very high frequency radio link. The controller used a CWT algorithm to calculate the train velocity and to activate the warning devices for 25 seconds, accordingly. A diagnostic remote monitoring system was later added to the grade crossing controller (Jordan, 2006).

After two successful trials, the system was installed at a passive crossing in Victoria for shadow mode testing. The equipment was powered by a solar generator. The active warning device, shown in Figure 8, consisted of a pair of flashing lights mounted on an Australian advanced warning sign installed 165–660 feet (50–200 meters) from both approaches to the crossing. As of 2006, the test results had not been published, but the authors stated that a revenue service test was forthcoming. The overall cost of the system was estimated to be in the order of one-fifth of the cost of a conventional active level crossing (Jordan, 2006).

During the trial period, several issues unrelated to the technology arose that influenced the selection of the electromagnetic induction loop technology: legal liability and the perception that government was trying to save money. "The legal advice received at the time took the view that the technology used had to be well tested, subjected to rigorous risk assessments, and applied in a professional manner. If this was done, it was concluded that a Court of Law would most likely have little reason to fine against the new device, all other matters being equal" (Government of Victoria, 2008).

Figure 8. Australian Advanced Warning Sign (Jordan, 2006)

4.   Current Research

In 2006, TC initiated a low-cost warning device research effort that was complementary to the research described in Section 3.4 above. The project began in 2006 and consists of two phases. The first, completed in 2007, comprised an analysis to identify potential solutions and an evaluation by a technical steering committee, resulting in selection of the two most promising technologies for testing. Phase 2, which is an ongoing effort, involves the acquisition, installation, and testing of the two candidate systems. The Canadian provinces of Quebec and Saskatchewan expressed interest in this research, and as such, a crossing in each province was selected for testing one of the low-cost warning systems (TC, 2007).

The two-phase process used by TC closely follows the methodology used by FRA/Volpe Center in the evaluation of alternative detection devices previously described. First, TC invited the North American grade crossing supplier industry to present its technologies, using the following general requirements:

- Cost between $10,000 and $30,000 Canadian
- Provide an optimal level of safety and reliability
- Operate independently of the railroad infrastructure with no onboard locomotive equipment or interconnection to the traditional railroad circuitry
- Be independent of the electrical grid
- Alert the locomotive crew to health of the grade crossing with sufficient advanced notice so that the crew can react appropriately and safely (TC, 2007)

Of the 20 companies that responded to the invitation, 11 were found by the steering committee to offer a system satisfying the specification and were invited to present their technology to the committee. As of this writing, only one system, manufactured by Carmanah Technologies Corporation, has been selected for testing. The train detection technology operates on the same physical principles as the EVA signaling product by measuring changes in the ambient magnetic field of the earth. However, instead of off-track magnetometers, a pair of triple-axis magnetoresistance detectors was installed at each approach to the crossing. Similarly, a pair was installed at each island entrance. The magenetoresistance detectors transmit occupancy information to the crossing control equipment by means of wireless radios. The warning system uses LED warning, and the entire crossing system is solar powered. If any of the detectors or radios fails, the warning devices are automatically activated (TC, 2007).

The minimum performance criteria for this system were defined by the steering committee as:

- Minimum 20 seconds of warning time
- Stop detection at the platform less than 2 seconds after train departure
- System operation over a speed regiment from 5 to 125 mph (8–200 km/h) (TC, 2007)

As with the FRA/Volpe Center alternative detection technology research, three types of failures were defined: *critical, missed detection,* and *nuisance/false alarms.* A critical failure was defined as a warning time of less than 20 seconds, system deactivation prior to train departure from the platform, and system deactivation greater than 10 seconds after train departure. Although a missed detection is self-explanatory, nuisance/false alarms consisted of any activation of the crossing warning system not resulting from a train (TC, 2007).

As of 2009, the Carmanah system was currently under evaluation at a private grade crossing in Saguenay, Quebec. A traditional active warning system was installed in parallel so as to compare the two technologies simultaneously. If the test results show that the system is highly reliable, TC will pursue a system failure analysis to evaluate how well it satisfies fail-safe operational principles.

5. Impediments to Acceptance

5.1 Institutional

As alluded to previously, tort liability remains the number one impediment to the implementation of low-cost warning devices at passive grade crossings. Two sections of the FRA Grade Crossing Signal System Safety regulation, 234.03 and 234.275, reinforce the complexity of the issue. Section 234.03 requires all grade crossing control circuits to operate on the fail-safe design principle. With over 100 years of field experience and millions of activation cycles, conventional train detection track circuits have been deemed essentially fail-safe by FRA, railroads, and suppliers. However, demonstrating that new technologies are fail-safe is difficult, especially at the low end of the cost spectrum. Low-cost warning technologies may be reliable, but low-cost, high-reliability, and fail-safe operation tend to be mutually exclusive features.

Section 234.75 requires non-track-circuit–based train detection technologies to comply with the safety and risk analysis requirements defined in 236, Subpart H, *Standards for Processor-Based Signal and Train Control Systems*. This is stated clearly in section 234.275 (e), where deviation from the fail-safe design requirement "...must be separately justified at the component, subsystem, and system level using the criteria of Section 236.909." Section 236.909, the minimum performance standard for products covered by Subpart H, requires that the introduction of such products does not result in risk that exceeds the previous condition. In the low-cost warning device domain, this means that the inherent risk associated with any new technology cannot surpass the risk of the preexisting passive crossing. The argument outlined above, notably the mutual exclusivity of low-cost, highly reliable, and fail-safe features, is equally applicable to proving that there is no increase in risk resulting from any new technologies.

The implication of these regulations is borne out by the reluctance of private or public sector entities to assume the liability associated with low-cost warning systems. As a result, the prevailing wisdom is that the tradeoff between cost and risk is too large and produces a negative cost-benefit curve.

The rational for Parts 234 and 236 most likely evolved from the legal discipline known as tort law. Tort law addresses, and provides remedies for, civil wrongs not arising out of contractual obligations. Although some categories of tort law deal with intentional wrongs, the relevant issue for grade crossings relates to unintentional wrongs or negligence. In this context, negligence is defined as conduct that falls below the standards of behavior established by law for the protection of others against unreasonable risk of harm. A person has acted "negligently" if he or she has departed from the conduct expected of a reasonably prudent person acting under similar circumstances. The hypothetical reasonable person provides an objective by which the conduct of others is judged. The key is that in tort law the reasonable person is not an average person or a typical person. Instead this person is a composite of what the community regards as to how the typical member should behave in situations that might pose a threat of harm to the public. In addition, the standard of conduct by which a person is judged is based on the activity or profession in which he or she is engaged. For example, the standard of conduct for a railroad signal engineer is compared with the conduct of a skilled, competent, and experienced railroad signal engineer (Ogden, 2007).

Liability for collisions occurring at grade crossings is governed by the law of negligence. Although tort law requires public agencies and railroads to exercise reasonable care to avoid injury to persons using the highway, they are not compelled to provide absolute safety. However, the potential for liability as a result of grade crossing collisions is a strong deterrent to testing and implementing new technology that deviates from conventional track circuit signaling systems (Ogden, 2007).

5.2 Cost

As described previously in this report, cost is one of the most difficult pieces of the low-cost warning device puzzle to solve. The primary reason is that low-cost, high-reliability, and fail-safe operation is almost always mutually exclusive properties. The procurement and installation costs must be sufficiently low to justify the investment. This is especially so at crossings with low-rail and highway traffic volumes. Although there is no established cost threshold, the research presented in this report suggests a range of 5–30 percent of the cost for a conventional track-circuit–based grade crossing system. The wide range in cost is attributable to the variation in performance requirements and the complexity of the failure analysis.

The labor expense associated with installation is typically 25–35 percent of the total system cost, as previously discussed in this report. This is not a trivial amount of money because it constitutes twice that of the train detection and warning device electronics. If, hypothetically, it were possible to completely eliminate the train detection and warning device expenses, the overall cost would still be 65–75 percent of a traditional grade crossing system.

Higher cost is also a consequence of the regulatory and legal barriers depicted earlier in this section. Decades of tort law precedent have produced a culture of risk aversion in the railroad industry. As a result, the industry is extremely reluctant to accept technologies that do not have a proven history of high reliability and safety. This is the current environment in which states or railroads are willing to invest large sums of money to reduce risk at grade crossings where the return on investment or benefit to society can be justified. However, the reality of limited funding precludes many passive public and private industrial crossings from receiving appropriate protection, either permanently or until the occurrence of a tragic accident.

5.3 Technological

Fail-safe operation is the overriding requirement for any active grade crossing system, regardless of the cost or application. Grade crossing train detection and warning systems are a subset of the larger family of safety-critical systems. Although many promising off-track and on-track low-cost grade crossing technologies have been evaluated, none has come close to satisfying the fail-safe operational requirement in the FRA Grade Crossing Signal System Safety regulations.

Many vendors claim that their system is designed in accordance with fail-safe operational principles. However, the definition of fail-safe is very specific and pertains to a design philosophy applied to safety-critical systems such that the result of hardware failure or the effect of software error shall either prohibit the system from assuming or maintaining an unsafe state or shall cause the system to assume a state known to be safe. Thus, demonstrating that a system is fail-safe is not trivial and can be costly.

The process of establishing that products designed solely from discrete mechanical or electrical components are fail-safe is usually rather straightforward. The probable failure modes of these systems can be analyzed exactly to ensure that a failure will not result in an unsafe condition. As such, these systems are referred to as intrinsically fail-safe. However, in recent years, processor-based technologies have gradually supplanted systems built solely from discrete components, and many contain a combination of both. This necessitates the use of a formal risk assessment methodology, which can be rather rigorous in nature. Although this methodology is also performed on discrete components, processor-based systems, which contain hardware and software components, are much more complex and require a much more thorough analysis. The primary steps in this process explained briefly below are summarized from 49 CFR Part 236, *Standards for Development and Use of Processor-Based Signal and Train Control Systems*, Appendix D:

- *Preliminary hazard analysis* (PHA) – follows or is performed in conjunction with the initial description of system requirements.
- *Hazard analysis* – focuses more on the detailed functions of the product and its components; an extension of the PHA performed in the later phases of product development.
- *Fault tree analysis* – identifies all hazards and determines their possible causes.
- *Failure modes and effects analysis* (FMEA) – considers the failure of any component within a system, tracks the effects of the failure, and determines its consequences. FMEA is particularly good at detecting conditions where a single failure can result in a dangerous situation. FMEA involves much detailed work and is expensive to apply to large complex systems.
- *Failure modes, effects, and criticality analysis* – an extension of FMEA that identifies the areas of greatest need.

The FRA grade crossing regulations are quite exacting in this respect, especially with regard to new or novel technologies as specified in 49 CFR 234 Section 234.275. This specifically applies to processor-based systems control systems not recognized for use prior to March 7, 2005, when 49 CFR Part 236 was issued, including non-track-circuit–based train detection and warning technologies. Many of the train detection technologies reviewed in this report are low cost in their commercial-off-the-shelf type. However, adding the appropriate safeguards to satisfy the safety regulations and reliability requirements for grade crossing and train control systems may negate any initial price advantage.

Reliability is frequently expressed in terms of mean time to failure. No detailed reliability values for grade crossing systems are specified in Federal regulations or railroad industry standards. Nevertheless, the railroad industry demands extremely high reliability from equipment that is critical to dependable revenue service operations. Although not necessarily a safety metric, reliability can be associated with safety under certain circumstances. For example, they are both components of mean time to hazardous event assessment, which is the measure of time likely to pass before the occurrence of a hazardous event.

Some of the basic technical concerns were presented in this section. However, the reality is that except for wayside wheel detector technology, none of the low-cost warning systems reviewed in this report approach the safety and reliability of track-circuit–based detection and warning systems. That being said, some are closer to implementation than others.

6. Findings

Conventional, track-circuit train detection and warning systems have a long history as the industry-accepted standard for grade crossing control systems. The superior performance and operational characteristics of these systems pose a real threat to the development of low-cost technologies that are non-track-circuit dependent. The challenge for the railroad supplier industry is to ensure that the safety and reliability of these non-track-circuit technologies are not compromised while still maintaining low cost.

This goal may be achievable because the term "low cost" is subjective and is highly dependent on a variety of factors. These include how the baseline grade crossing warning device cost is defined, the cost threshold defined by the user, the requirements specific to the application, and the complexity of the application. Consider that the technologies described in this report ranged from 10 to 30% of the baseline cost. Installation of grade crossing infrastructure is the largest single-cost driver for traditional grade crossing systems. This includes trenching and connecting to the local utility grid, which can be expensive if no connection is available on-site. This necessitates a robust solar power system capable of supporting the train detection and warning device equipment.

The wayside wheel detector technology has shown the most promise and is probably closest to achieving acceptance by the railroad industry. This does not imply, however, that the other technologies are not potentially viable with additional research and development. For one thing, the wayside wheel detector system may not ultimately be the lowest-cost technology, especially over the long term. In addition, the objective of this research was to present an objective assessment of the available low-cost warning device technologies and recommend a migration path that would facilitate their implementation.

The technologies described in this report fall into one of two categories—on-ROW or off-ROW. Although off-ROW systems are less problematic to install, they tend to be less safe and reliable than those that are on-ROW. The wheel detector is an on-ROW technology that has already been shown to be fail-safe and vital.

The institutional impediments to acceptance of low-cost grade crossing warning devices have proven to be the most intractable. Cost control, which cuts across all aspects of this research, is the underlying theme in this report. Reducing cost is a fundamental requirement, because low-cost, high-reliability, and inherent fail-safe operation are as a rule mutually exclusive. However, the reader should keep in mind that locations where external factors such as highway traffic, rail traffic, and/or accident frequency is intolerably high, cost is not a factor. The focus of low-cost warning device research is to lower the risk at passive grade crossings that are marginally risky, and the installation of conventional warning devices cannot be justified on safety alone.

Although not likely, there is the possibility of a technological breakthrough resulting in a reduced cost structure for these systems and a high level of reliability and safety. As research in this field is always evolving, this outcome cannot be ruled out. However, as pointed out earlier, labor and installation costs tend be rather fixed. Also, the attempts to find the so-called low-cost/high-reliability and safety "sweet spot" have been unsuccessful.

Cost is an all-encompassing element that dovetails directly into the other institutional impediments—technology acceptance and reliability/safety. The precedent in the tort law system in the United States has been long established, and there is little likelihood for change in the foreseeable future. Although not perfect, conventional track circuits are the accepted technology for train detection and warning systems at grade crossings. If a passive grade crossing is upgraded with non-track-circuit–based technology, the new system must at a minimum be at least as safe as conventional track circuits. In addition, when a passive crossing is upgraded with active devices, the burden of responsibility for train detection transfers from the motor vehicle user of the crossing to the railroad. In light of these two concepts, there is little incentive for the railroad industry to innovate and a great deal of downside risk. Thus, the same modern-day legal system that requires safety-related industries to design and build a system to the most stringent safety requirements also lacks the flexibility to allow the introduction of many new, potentially life-saving technologies.

The final piece of the puzzle is demonstrating that a new technology is fail-safe and reliable. Historically, the safety assurance logic for track-circuit–based train detection and warning has resided in electromechanical relays. These devices are renowned for simple design, highly reliable operation, and few but very well understood failure modes. In recent years, relays have been supplanted by processor-based systems, with similar underlying logic. FRA has monitored these systems for years and has incrementally increased its confidence that these systems are indeed safe and reliable, thus allowing many of theme systems to be "grandfathered-in."

By contrast, the low-cost warning device technologies in this research deviate significantly from traditional track circuit design and operation. Although the designs of these systems vary in complexity, the failure modes are not well understood. Also, FRA regulations, most notably 49 CFR 234.275, specify well-defined performance standards for new non-track-circuit–based technologies. Although a coherent approval process is now in place, the level of effort required by the railroad supplier industry to show that these technologies are safe and reliable probably outweighs the potential monetary benefits.

7. Recommendations

Leverage Regulatory Efforts to Break Down Institutional Barriers
The chance that technological innovation will create the "magic bullet" that provides the industry with a low-cost, highly safe, and reliable grade crossing technology is an unlikely prospect. One alternative is to reduce the legal and regulatory burden on railroads and railroad suppliers. The solution may reside in an amendment to Title 49 of the *U.S. Code* known as the Rail Safety Improvement Act (RSIA) of 2008, signed into law by President George W. Bush in October 2008. Section 210 of the RSIA, *Fostering Introduction of New Technology to Improve Safety at Highway-Rail Grade Crossings*, supplies regulatory relief to railroads and suppliers. The authors of the amendment recognize that "the emergence of new technologies" may potentially encourage "more effective and affordable warnings," resulting in increased safety for highway users and trains at passive grade crossings. In addition, they acknowledge that implementation of new grade crossing technology will not be easy and will require substantial cooperation between authorities and railroads to ensure success. The technical guidance in Section 210 provides reference to 49 CFR Part 236 Subpart H, *Standards for Development and Use of Processor-Based Signal and Train Control Systems*. The authors clearly describe this regulation as providing a "suitable framework" for qualifying "new or novel" grade crossing technology.

The specific language that may provide regulatory relief is as follows:

"If the Secretary approves by order new technology to provide warning to highway users at a highway-rail grade crossing and such technology is installed at a highway-rail grade crossing in accordance with the conditions of the approval, this determination preempts any State statute or regulation concerning the adequacy of the technology in providing warning at the crossing."

Of course, this is not a total panacea, and the regulatory burden is by no means insignificant. However, the Secretary of Transportation now has the authority to approve new technology. This authority supersedes any State laws, potentially resulting in a lower tort liability risk. Also, the use of performance-based standards for technology qualification relieves vendors from the burden of having to prove that the products are "fail-safe."

Another option may be to change the perception of non-fail-safe active warning devices in the United States. In Australia, there is a movement to implement these systems as supplemental or enhancing existing signage at passive grade crossings. This would put them at a level equivalent to highway traffic control devices. In this manner, highway users of the grade crossing would still be responsible for train detection. Currently, the Australian Department of Transport is evaluating the potential necessity for developing regulatory language to introduce non-fail-safe technology, especially where passive grade crossings are located. This dovetails well with the requirements specified in Section 210 of the RSIA of 2008.

Engage in Cooperative Research with TC
Both FRA and TC are engaged in similar low-cost warning device research. A collaborative research effort could afford multiple opportunities to leverage costs, resources, and institutional knowledge. This research could result in a common North American low-cost warning performance criteria (see Appendix A for a sample) and performance guidelines for different operating environments (i.e., dark versus signalized territory and different levels of signaling scenarios). At a minimum, FRA should continue monitoring the progression of the TC research and document the lessons learned. The triple-axis magnet resistance train detection technology, if successfully tested, is potentially a "true" low-cost solution and may be a candidate for qualification through the 2008 RSIA.

Human Factors Simulation Evaluation of Potential Low-Cost Technologies
Under previous FRA research, the Volpe Center developed a driving simulator to evaluate driver response to different train reflector patterns. This desktop simulator included a personal computer for generating the vehicle dynamics and visual scenery, a steering wheel, pedal controls, and driving dynamics to simulate a vehicle traveling on a typical two-way American rural road. This simulator has subsequently been upgraded and could be modified to evaluate driver response to different low-cost warning device technologies and signage at and approaching grade crossings. This will ensure a broad survey of available technologies. Based on the results of this evaluation, the best performing technologies could be ranked and identified for possible field testing.

Evaluate Potential Implications of Warning Device and Sign Changes
The MUTCD, published by the Federal Highway Administration (FHWA), is the national standard for all traffic control devices installed on any street, highway, or bicycle trail open to public travel. The MUTCD is specifically approved by FHWA and requires that each State be in conformance with the standards set forth within the chapters. All additions, revisions, or changes to the MUTCD are developed by the National Committee on Uniform Traffic Control Devices (NCUTCD), a private, nonprofit organization whose purpose is to assist in the development of the MUTCD. NCUTCD is supported by 21 sponsoring organizations, including AAR.

Part 8 of the MUTCD, *Traffic Controls for Highway-Rail Grade Crossings*, describes traffic control devices, systems, and practices for grade crossings open to public travel. The standards are extremely specific and were developed through consensus by the NCUTCD membership. These include physical dimensions, graphics and colors, and locations of grade crossing warning signs, gates, and lights. Any modifications or additions to these standards resulting from low-cost warning device application will need to be vetted by the NCUTCD and incorporated into the MUTCD.

Test Promising Technologies Identified by Previous Research
Numerous non-track-circuit–based concepts have been documented in this report, ranging in maturity from conceptual to approaching commercial viability. The two most promising train detection technologies are based on GPS (e.g., Minnesota) and magnetic

flux (e.g., wheel sensors). Like the rest of the technologies reviewed in this report, each of these presents unique opportunities and challenges. For example, the Mn/DOT system locomotive-centric train detection system eliminates the need for installing and maintaining more vulnerable wayside detection systems. However, the designers were unable to satisfy the fail-safe operational requirement in 49 CFR Part 234.203 of the FRA grade crossing regulation. However, Section 210 of the 2008 RSIA presents an unusual chance to qualify this technology for certain rural operating environments. The wheel sensor technology has a fail-safe design but has yet to garner the interest from the North American railroad industry to reach the threshold of commercial viability. Also, the cost level of this technology is not well understood.

References

Chuang, S.L, Hsu, A., & Young, E. (2003). *Fiber Optical Sensors for High-Speed Rail Applications*. Final Report for High-Speed Rail IDEA Project 19. University of Illinois at Urbana-Champaign, Urbana, IL. Prepared for Transportation Research Board of the National Academies, Washington, DC.

Federal Railroad Administration. (2002 June 5). Petition for Waiver of Compliance: Twin Cities and Western Railroad Company. *Federal Register*, 67, 38695.

Federal Railroad Administration. (2009a). *Railroad Safety Statistics 2007: Final Annual Report*. Washington DC: U.S. Department of Transportation.

Federal Railroad Administration. (2009b). *Highway-Rail Crossing Program*. Retrieved July 29, 2009, from http://www.fra.dot.gov/us/content/86.

General Accounting Office. (1995). *Railroad Safety: Status of Efforts to Improve Railroad Crossing Safety*. Washington DC.

Government of Victoria. (2008). *Inquiry into Improving Safety at Level Crossings*. Melbourne, Australia.

Guins, T.S., Reiff, R.P., Gurule, S., & Gage, S. (1996 June). *Results of Alternative Train Presence Detection Systems Tests*. Association of American Railroad Technology Digest. Washington DC.

Hildebrand, E., Roberts, C., & Robichaud, K. (2007). *A Low-Cost Rail Warning System for Private and Farm Road Crossings*. University of New Brunswick Transportation Group. Prepared for Transportation Development Centre and ITS Office of Transport Canada and New Brunswick Department of Transportation.

Jennings, S.L., Field, J., Worley, P., & Scott, P. (2005 April). *EVA Signal System Report*. Prepared for North Carolina Department of Transportation, Raleigh, NC.

Jordan, P. (2006, February). A Trial of a Low-Cost Level Crossing Warning Device. *The Institution of Railway Signal Engineers, Australasian Section*, Issue 1, pp. 1–5.

Lerner, N.D., Llaneras, R.E., McGee, H.W., & Stephens, D.E. (2002). *NCHRP Report 470: Traffic-Control Devices for Passive Railroad-Highway Grade Crossings*. National Academy Press, Washington, D.C. Transportation Research Board — National Research Council.

Luedeke, J.F., Wagner, D.P., Carroll, A.A., & Markos, S.H. Safety of Highway-Railroad Grade Crossings: *Volume I: International Signaling and Control Technologies for High-Speed Rail Applications*. Unpublished manuscript.

Millegan, H., Yan, X., Richards, S., & Han, L. (2009 January). *Evaluation of Effectiveness of Stop-Sign Treatment at Highway-Railroad Grade Crossings.* The Center for Transportation Research, The University of Tennessee, Knoxville. Presented at the TRB 88th Annual Meeting, Washington, DC.

Moody, H., & Reiff, R. (2001 November). *Performance Requirements for Grade Crossing Warning Systems.* Presented at the Town Hall Meeting on Low Cost Options for Grade Crossing Systems, 2001 National Highway-Rail Grade Crossing Safety Conference, Dallas, TX.

National Transportation Safety Board. (1998). *Safety At Passive Grade Crossings. Volume 1: Analysis. Safety Study.* NTSB/SS-98/02. Washington, DC.

Ngamdung, T. (2009a). [Incident Rate at Public Active and Passive Crossing, 1989–2008]. Unpublished raw data.

Ngamdung, T. (2009b). *Strategic Historical Data Analysis of Grade Crossing and Trespass Incidents.* Unpublished presentation.

O'Conner, J. (2009 January). Microwave Train Detection System for Grade Crossings. *New Ideas for High Speed Rail*, pp. 23–24. Transportation Research Board of the National Academies, Washington, DC.

Ogden, B.D. (2007). *Railroad-Highway Grade Crossing Handbook - Revised Second Edition 2007.* Prepared by Institute of Transportation Engineers. Federal Highway Administration. Washington DC: U.S. Department of Transportation.

Olson, L.E., & Roop, S.S. (2003). *An Investigation into the Use of Buried Fiber Optic Filament to Detect Trains and Broken Rail.* Final Report for High-Speed Rail IDEA Project 18. Texas Transportation Institute, College Station, TX. Prepared for Transportation Research Board of the National Academies, Washington, DC.

Peterson, B. (2001 December 1). The $10,000 Light Bulb. *Railway Track and Structures*, 97, 11-13.

Petit, B. (2002 May 1). The Previously Untold Story About Grade Crossing Costs. *Railway Track and Structures*, 98, 14–15.

Reiff, R.P., Gage, S.E., Caroll, A.A., & Gordon, J.E. (2003). *Evaluation of Alternative Detection Technologies for Trains and Highway Vehicles at Highway Rail Intersections.* Washington DC: Federal Railroad Administration, U.S. Department of Transportation. DOT/FRA/ORD-03/04, DOT-VNTSC-FRA-03-02.

Roop, S.S., Roco, C.E, Olson, L.E., & Zimmer, R.A. (2005). An Analysis of Low-Cost Active Warning Devices for Highway-Rail Grade Crossings. Texas Transportation

Institute, The Texas A&M University System, College Station, Texas. Prepared for National Cooperative Highway Research Program, Transportation Research Board, National Research Council, Washington, DC. NCHR Project No. HR 3-76B

Russell, E., Rys, M., & Liu, L. (1999). *Low-Volume Roads and the Grade Crossing Problem*. Transportation Research Record 1652, pp. 79–85. National Research Council, Washington, DC.

Transport Canada. (2007). *Identification and Evaluation of Low-Cost Warning Systems at Grade Crossings: Draft Project Implementation Plan*. Transportation Development Center, unpublished manuscript.

URS Corporation & TranSmart Technologies, Inc. (2005). *Low-Cost Highway-Rail Intersection Active Warning System Field Operational Test: Evaluation Report*. Prepared for Minnesota Department of Transportation Office of Traffic, Security and Operations, St. Paul, MN.

Appendix A.
Sample Performance Criteria

Table A-1. Sample Performance Criteria from TTI Analysis of Low-Cost Warning Devices (Roop, Roco, Olson, and Zimmer, 2005)

Category	Criteria	Description
Enhance Safety	Fail-Safe Design	Yes-System detects internal component failures and activates the warning device No-System does not activate the warning device in all instances of component failure
	Provides 20-Second Advanced Warning	Yes-System activates the warning device in compliance with the 20-second rule No-Does not comply with "Yes" as described above
	Geometric Flexibility	Yes-The recommended system configuration is capable of detecting the presence of trains through or around obstacles (i.e., curves, trees, buildings, etc.) No-Does not comply with "Yes" as described above
Reduce System Cost	Installation Cost	Dollar value associated with initial construction, activation, testing, and calibration High-More than $8,000 Medium-$2,000 to less than $8,000 Low-Less than $2,000
	System Life	Projected lifespan of the proposed detection and control system (excludes the assumed standard warning device; i.e., pole with cross bucks and flashing lights): High-More than 20 years Medium-10 years to less than 20 years Low-Less than 10 years

Table A-1. Sample Performance Criteria from TTI Analysis of Low-Cost Warning Devices (Roop, Roco, Olson, and Zimmer, 2005)

Category	Criteria	Description
	Failure Rate	Low-Components comply with U.S. military specifications quality Medium-Components comply with industrial-grade quality High-Components are obtained from surplus sources
	Redundancy	Yes-All system elements are duplicated No-Does not comply with "Yes" as described above
Reliability	Simplicity	Simple-Uses a single electromechanical sensor, sensor and control system are contained in the same enclosure, sensor and control system enclosure are mounted on the warning device Marginal-Uses multiple electromechanical sensors, sensor and control system are contained in the same enclosure, sensor and control system enclosure are mounted on the warning device Moderate-Uses multiple electromechanical sensors, sensor and control system are enclosed separately and require a hardwire connection Complex-Uses existing or comparable track circuitry. Requires direct contact with the track Very Complex-Requires multiple sensors and activation control points that share communication (typically via radio)
	Disruption Resistance	High-Control system is contained in a buried, hermetically sealed vault Good-Control system is locked in an aboveground steel or concrete bungalow Moderate-Control system is locked in an aboveground cabinet Poor-Control system is contained in an unlocked, aboveground cabinet
	Known/Controlled Failure Modes	Yes-Failure modes have been identified and are controlled No-Does not comply with "Yes" as described above

Table A-1. Sample Performance Criteria from TTI Analysis of Low-Cost Warning Devices
(Roop, Roco, Olson, and Zimmer, 2005)

Category	Criteria	Description
Installability	Ease of Installation	Easy-Requires installation of a single component (i.e., the warning device with mast-mounted detection, control, and power supply equipment) Moderate-Requires installation of a warning device, a separately enclosed control system, and a hardwire connection between the warning device and control system Hard-Requires installation of a warning device, a control system in a locked vault or bungalow, multiple distributed sensors with an access road, and hardwire connection between the warning device, control system and sensors
	Site Preparation	Low-Preparation is restricted to that required for installation of the warning device Medium-Requires construction of a simple access roadway High-Requires modification of the existing terrain through earthwork, removal of site vegetation
Maintainability	Ease of Inspection & Testing	Easy-Requires a simple volt-ohm-meter (VOM) to evaluate system compliance with voltage and resistance operating requirements that can be performed with minimal training Moderate-Requires VOM and current-meter to evaluate system compliance and requires a moderately skilled electronic technician to test the system Hard-Requires test equipment specifically designed for the system and a certified technician to test the system
	Ease of Repair	Easy-Replacement of components requires simple pull-out/plug-in repair Moderate-Replacement of components requires multiple reconnections of components Hard-Requires system recalibration following the replacement of each component
Compatibility	Encroaches on Railroad Property	No-Does not rely on components of the existing detection railroad infrastructure as elements of the train system Yes-Does not comply with "No" as described above

Abbreviations and Acronyms

AAR	Association of American Railroads
AREMA	American Railway Engineering and Maintenance-of-Way Association
ATV	all-terrain vehicle
BNSF	Burlington Northern Santa Fe
CFR	Code of Federal Regulations
CWT	constant warning time
FHWA	Federal Highway Administration
FMEA	failure modes and effects analysis
FOT	field operational test
FRA	Federal Railroad Administration
GPS	global positioning system
IDEA	Innovations Deserving Exploratory Analysis
ITS	intelligent transportation system
km	kilometer(s)
km/h	kilometer(s) per hour
LED	light-emitting diode
Mn/DOT	Minnesota Department of Transportation
MOW	maintenance-of-way
mph	mile(s) per hour
MUTCD	*Manual of Uniform Traffic Control Devices*
NCDOT	North Carolina Department of Transportation
NCUTCD	National Committee on Uniform Traffic Control Devices
NCVARR	North Carolina and Virginia Railroad
PHA	preliminary hazard analysis
RFTI	Request for Technical Information
ROW	right-of-way
RSIA	Rail Safety Improvement Act
T-M	traffic moment
TC	Transport Canada
TC&W	Twin Cities and Western Railroad
TRB	Transportation Research Board
TTC	Transportation Technology Center
TTCI	Transportation Technology Center, Inc.
TTI	Texas Transportation Institute
Volpe Center	John A. Volpe National Transportation Systems Center
VOM	volt-ohm-meter

www.ingramcontent.com/pod-product-compliance
Lightning Source LLC
Chambersburg PA
CBHW081758170526
45167CB00008B/3239